JUEL ANDERSEN'S SEA GREEN PRIMER

BY

RICHARD FORD
JUEL ANDERSEN

OTHER BOOKS

BY JUEL ANDERSEN

THE TOFU COOKBOOK *WITH CATHY BAUER*
TOFU KITCHEN
TOFU PRIMER
TOFU FANTASIES
TEMPEH PRIMER *WITH ROBIN CLUTE*
CAROB PRIMER *WITH ROBIN CLUTE*

BY RICHARD FORD

THE SOY FOODERY COOKBOOK

JUEL ANDERSEN'S

SEA GREEN PRIMER

A BEGINNER'S BOOK OF SEA WEED COOKERY

by RICHARD FORD

JUEL ANDERSEN

with SIGRID ANDERSEN

Illustrations from *Little Pictures of Japan;* Edited by Olive Beaupre Miller with pictures by Katharine Sturges, The Book House for Children, Chicago, 1925.

Book design and lettering by Sigrid Andersen

ISBN 0-916870-65-0

Published by:
CREATIVE ARTS BOOK COMPANY
833 Bancroft Way
Berkeley, California 94710

CONTENTS

INTRODUCING SEA GREENS

SEA GREENS ARE PLANTS OF THE OCEANS. WE AMERICANS CALL THEM "SEAWEEDS" AND THINK OF THEM AS THINGS THAT WASH UP ON OCEAN BEACHES DURING STORMS OR AS A BED FOR COOKING SEAFOODS AT CLAMBAKES. YET THESE WEEDS, OR VEGETABLES FROM THE SEA, ARE AN IMPORTANT FOOD FOR WE WESTERN PEOPLES AND A MAJOR FOOD FOR THE PEOPLE OF THE ORIENT, POLYNESIA, INDONESIA, AND THE PHILLIPINES.

SEA GREENS COME IN MANY SIZES, SHAPES, AND COLORS. THEY RANGE FROM THE MICROSCOPIC TO THOSE OVER ONE HUNDRED FEET LONG. SOME ARE FLAT, SOME ARE ROUND, OTHERS FEATHERY OR ELONGATED. THERE ARE THOUSANDS OF KNOWN SPECIES WITH A WIDE RANGE OF COLORS: BLUES, GREENS, REDS, AND BROWNS.

BECAUSE THEIR CARBOHYDRATE IS MOSTLY INDIGESTIBLE, THEY PROVIDE MUCH NUTRITION, BUT FEW CALORIES. IN THE DRY FORM, SEA GREENS ARE ABOUT 25% PROTEIN AND A LOW 2% FAT.

THEY CONTAIN THE IMPORTANT MINERALS SODIUM, POTASSIUM, AND CALCIUM IN ALMOST THE SAME PROPORTION AS SEA WATER AND OUR OWN BODY TISSUES. THEY ARE SOURCES OF IODINE, IRON, MAGNESIUM, ZINC, AND PHOSPHORUS AND CONTAIN AN ARRAY OF VITAMINS: A, B_1, B_2, B_6, B_{12}, AND C, PLUS PANTOTHENIC ACID, FOLIC ACID, AND NIACIN IN RICHER SUPPLY THAN LAND VEGETABLES.

UNKNOWN TO MOST OF US, WE HAVE EATEN SEA GREEN PRODUCTS ALL OUR LIVES DISGUISED AS AN INGREDIENT IN PROCESSED CHEESES, BAKED GOODS, ICE CREAMS AND MILKS, YOGURTS, DAIRY PREPARATIONS, CONDIMENTS, JAMS AND JELLIES, PROCESSED MEATS, AND A HOST OF OTHER COMMON FOODS. WE HAVE ALL SEEN THE WORDS "CARRAGEENAN", "ALGIN", AND "AGAR" TIME AND AGAIN; THESE ARE SEA PLANT PRODUCTS.

IN ADDITION THEY ARE WIDELY USED IN NON-FOOD PREPARATIONS: MEDICINES, COSMETICS, INDUSTRIAL PRODUCTS AND FERTILIZERS, AND AS AN IMPORTANT SOURCE OF SODA, POTASSIUM, AND IODINE.

IN THEIR NATURAL FORM, AS AN ACTUAL FOOD, THESE VEGETABLES FROM THE SEA ARE JUST COMING INTO USE IN AMERICAN STYLE DISHES. WE HAVE CHOSEN TO CALL THEM "GREENS" IN THIS BEGINNER'S COOKERY BOOK BECAUSE THEY OCCUPY A POSITION IN OUR DIET MOST CLOSELY ALLIED TO OTHER MORE FAMILIAR GREENS, SUCH AS SPINACH, CHARD, AND LETTUCE. BUT THE USES OF "SEA" GREENS ARE WIDER THAN THOSE OF "LAND" GREENS, AS YOU WILL SOON LEARN.

THOUGH SEA GREENS ARE USED IN MANY COUNTRIES, THEIR USE IN JAPAN HAS BEEN THE MOST WIDE. THE SEAWEED INDUSTRY THERE DATES BACK TO THE 14TH CENTURY; THE ACTUAL FARMING OF SEA PLANTS BEGAN IN THE 1700'S IN SHALLOW COASTAL WATERS. PROBABLY 10% OF THE JAPANESE DIET CONSISTS OF SOME FORM OF SEA GREEN.

WE HAVE CHOSEN SEVEN OF THE MOST COMMONLY AVAILABLE PACKAGED SEA GREENS TO EXPLORE. THEY ARE BEST KNOWN THIS COUNTRY BY THE FAMILIAR NAMES: AGAR AND DULSE, AND THE LESS FAMILIAR JAPANESE NAMES: ARAME, HIJIKI, KOMBU, NORI, AND WAKAME.

STORING SEA GREENS

SEA GREENS CAN BE PURCHASED IN ALL ORIENTAL FOOD STORES AND IN MOST NATURAL OR HEALTH FOOD STORES. THEY ARE SOLD IN SEALED PACKAGES OR IN BULK IN THE DRY FORM. THEY SHOULD BE STORED IN A DRY, DARK PLACE WHERE THE TEMPERATURE IS LOW. HIGH TEMPERATURE, HIGH HUMIDITY, AND PROLONGED STORAGE WILL CAUSE SOME BREAKDOWN OF VITAMINS A AND C, ALTHOUGH SOME DRIED SEA GREENS CAN BE STORED FOR YEARS. IF PLANTS PICK UP MOISTURE, DURING STORAGE, PLACE THEM IN AN OVEN AT A LOW TEMPERATURE (100°) UNTIL CRISPNESS IS RESTORED.

GLOSSARY

MISO ~ FERMENTED SOY BEAN PASTE, USED AS A FLAVORING

MOCHI ~ PRESSED SWEET RICE PRODUCT USED AS A BREAD

TAHINI ~ GROUND SESAME SEEDS, SOMETIMES CALLED SESAME BUTTER

TAMARI ~ A HIGH SODIUM, FERMENTED CONDIMENT ALSO CALLED SHOYU AND SOY SAUCE MADE FROM SOYBEANS AND VARIOUS GRAINS

TEMPEH ~ CULTURED WHOLE SOYBEAN AND GRAIN FOOD, NATIVE TO INDONESIA

TOFU ~ OR DOUFU, SOY BEAN CURD MADE FROM SOY MILK

SURIBACHI ~ AN EARTHENWARE BOWL, SCORED AT THE BOTTOM AND USED AS A MORTAR

AGAR

GELIDIUM AMANSI
"AGAR", "KANTEN"
"TENGUSA"

AGAR IS GELATIN FROM SEAWEED. IT IS MADE FROM THE MANY TYPES OF RED ALGAE THAT GROW IN MOST OF THE WORLD'S SEAS. IT IS THE MOST IMPORTANT AND WIDELY USED SEA GREEN, FROM WHICH WE GET ALGIN, CARRAGEENAN, AND THE AGAR WE WILL LEARN TO USE. ALGIN AND CARRAGEENAN HAVE WIDE COMMERCIAL USES BUT ARE NOT AS READILY AVAILABLE FOR HOME USE AS AGAR.

THE JAPANESE HAVE USED AGAR FOR CENTURIES, AND ACCORDING TO LEGEND, DISCOVERED THE WAY TO PURIFY IT BY ACCIDENTLY FREEZING IT. IT WAS CALLED KANTEN WHICH MEANS "COLD SKY".

AGAR, OR "AGAR-AGAR", AS AMERICANS CALL IT, IS SOLD IN FLAKES AND POWDER FORM, ALTHOUGH IT CAN ALSO BE PURCHASED IN BARS OR STICKS CALLED "KANTEN". IT IS COLORLESS, ODORLESS, TASTELESS, AND STORES WELL. IT DISSOLVES EASILY WITHOUT LUMPING AND JELLS AT JUST UNDER 100° F., REMAINING STABLE, WITHOUT REFRIGERATION.

CARRAGEENAN IS MADE FROM IRISH MOSS AND IS NAMED FOR AN IRISH VILLAGE, CARRAGHEEN. IT IS USED COMMERCIALLY AS A STABILIZER IN DOZENS OF FOOD PRODUCTS. ALGIN, A CLOSE COUSIN, IS USED AS AN EMULSIFIER AND STABILIZER IN BOTH FOOD AND NON-FOOD PRODUCTS. ALL ARE USED IN COSMETICS, MEDICINES, AND INDUSTRIAL PRODUCTS.

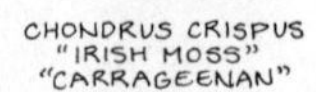

CHONDRUS CRISPUS
"IRISH MOSS"
"CARRAGEENAN"

AGAR, CONTINUED

NUTRIENTS: AGAR IS RICH IN CALCIUM, PHOSPHORUS, IODINE, AND BROMINE AS WELL AS IMPORTANT TRACE MINERALS. ITS CALORIE COUNT IS NEGLIGIBLE; IT PROVIDES BULK FOR DIETERS, QUELLING HUNGER WHILE PROVIDING NUTRIENTS.

PREPARATION: AGAR IS EASY TO WORK WITH. ONE EXPERIENCE WITH IT WILL CAUSE THE USER TO FOREVER ABANDON ANIMAL GELATIN, WHICH LUMPS AND HAS A NOXIOUS ODOR. IT CAN BE MELTED AND GELLED AGAIN, WHICH CAN BE MARVELOUS IN MAKING LAYERED DISHES. USE IN THE FOLLOWING PROPORTIONS TO REPLACE GELATIN IN ANY RECIPE:

TO JELL 2 CUPS OF LIQUID:

2 TEASPOONS AGAR POWDER ≡ 4 TEASPOONS GELATIN *

3 TBSP. AGAR FLAKES ≡ 4 TEASPOONS GELATIN

1 BAR KANTEN ≡ 4 TEASPOONS GELATIN

* 4 TEASPOONS GELATIN = 1 ENVELOPE

GELIDIUM LATIFOLIUM "AGAR"

AGAR NEED NOT BE SOAKED IN COLD WATER AS IS NECESSARY WITH GELATIN. ADD IT TO LIQUID, BRING TO A BOIL AND STIR UNTIL IT IS COMPLETELY DISSOLVED. KANTEN, HOWEVER, SHOULD BE SHREDDED AND SOAKED, AND WILL TAKE LONGER TO MELT, ALTHOUGH IT IS USED IN THE SAME WAY AS AGAR.

USES: FOR DESSERTS, JELLIES, LIGHT PUDDINGS, ASPICS, SALADS, UNCOOKED CHEESE CAKES, PATÉS, PRESERVES, CONDIMENTS, SOUPS, FILLINGS, ICE CREAM OR MILK, ETC.

SIMPLY CHEESECAKE

SERVES 8

CRUST

1 CUP CEREAL, COOKIE, OR BREAD, CRUMBS
1/4 CUP MELTED BUTTER
1 TBSP. HONEY

FILLING

1 1/2 CUPS MILK OR SOYMILK
1/2 CUP HONEY
2 TEASPOONS AGAR POWDER OR
1/3 CUP AGAR FLAKES
14 OZ. FIRM TOFU
1/4 LB. STRONG CHEDDAR CHEESE, AT ROOM TEMPERATURE
1/2 TEASPOON SALT
1/2 CUP LEMON JUICE
2 TEASPOONS GRATED LEMON RIND

TOPPING

1 CUP FRESH FRUIT: STRAWBERRIES, RASPBERRIES, BLUEBERRIES, ETC. *
1 CUP APPLE JUICE
• HONEY OR SUGAR TO TASTE
1/2 TEASPOON AGAR POWDER OR 4 TEASPOONS AGAR FLAKES

PREPARING CRUST

PREPARE A CRUMB CRUST BY MELTING THE BUTTER (OR ANY SHORTENING) AND ADDING THE HONEY TO IT. COMBINE WITH THE CRUMBS IN A SMALL BOWL AND MIX WELL. PAT INTO THE BOTTOM OF AN 8" SPRING

(CONTINUED NEXT PAGE)

SIMPLY CHEESECAKE, CONTINUED

FORM OR PIE PAN AND REFRIGERATE WHILE PREPARING THE FILLING. THIS CRUST NEED NOT BE BAKED, BUT BAKING WILL MAKE IT VERY CRISP. (BAKE AT 350° FOR 10 MINUTES.)

PREPARING FILLING

COMBINE THE MILK, HONEY, AND AGAR IN A SAUCEPAN AND BRING TO A BOIL, STIRRING UNTIL THE AGAR AND HONEY ARE COMPLETELY MELTED. COMBINE THE TOFU, CHEESE, SALT, LEMON JUICE, AND LEMON RIND IN A BLENDER OR FOOD PROCESSOR AND BLEND UNTIL VERY SMOOTH. ADD THE HOT MILK MIXTURE AND BEAT WELL. POUR INTO A MIXING BOWL AND REFRIGERATE UNTIL SET. WHEN SET, BEAT WITH A ROTARY BEATER UNTIL LIGHT AND FROTHY. POUR INTO THE PREPARED CRUST AND REFRIGERATE UNTIL CHILLED.

PREPARING TOPPING

CLEAN AND PREPARE THE FRUIT BY CUTTING OR WHATEVER IS NECESSARY. COMBINE THE FRUIT JUICE, SWEETENER, AND AGAR IN A SAUCEPAN AND HEAT UNTIL MELTED. COOL, BUT DO NOT ALLOW IT TO SET. SPREAD THE FRUIT OVER THE CHEESECAKE IN ANY MANNER YOU LIKE. (A PATTERN ADDS AN ELEGANT TOUCH.) THEN POUR THE STILL WARM MIXTURE OVER THE TOP, BEING CAREFUL TO COVER ALL OF THE FRUIT TO PRESERVE IT AND KEEP IT FROM RUNNING. CHILL BEFORE SERVING.

* FROZEN FRUIT WORKS WELL TOO. THAW AND DRAIN IT, AND USE THE JUICE BY ADDING WATER TO MAKE 1 CUP. PROCEED AS WITH FRESH FRUIT.

TOMATO ASPIC

SERVES 4

4 CUPS TOMATO JUICE AND WATER, IN ANY PROPORTION
~ FLAVORING TO TASTE: WORCESTERSHIRE, LEMON JUICE, HOT SAUCE, OR SAUERKRAUT JUICE, ETC.
2 TEASPOONS AGAR POWDER OR 2 TBSP. AGAR FLAKES

COMBINE TOMATO JUICE, WATER, AND FLAVORING IN A SAUCEPAN AND TASTE BEFORE HEATING. ADD THE AGAR, STIR WELL, AND HEAT TO BOILING, STIRRING UNTIL AGAR IS MELTED. POUR INTO A MOLD AND COOL TO ROOM TEMPERATURE BEFORE REFRIGERATING.

THIS MAKES A LOVELY BASE FOR ANY NUMBER OF INNOVATIVE SALADS. THE INGREDIENTS YOU CHOOSE TO ADD CAN BE FOLDED IN BEFORE REFRIGERATING. USE CUT OR CHOPPED VEGETABLES, FISH OR MEATS, CHEESES, OR ANYTHING YOU LIKE. CHILL BEFORE SERVING.

VEGETABLE ASPIC

SERVES 3 TO 4

2 CUPS CLEAR SOUP STOCK (KOMBU STOCK, PAGE 42)
2 TBSP. AGAR FLAKES OR 2 TSP. AGAR POWDER
2 TBSP. TAMARI
1 BEET, CHOPPED
1 STALK CELERY, SLICED VERY THIN
1 CARROT, DICED
1 LB. FRESH PEAS, SHELLED
1 SMALL HEAD CABBAGE, FINELY SLICED
1/2 BUNCH SCALLIONS, FINELY CHOPPED

BRING STOCK, AGAR, AND TAMARI TO A BOIL; REDUCE HEAT AND SIMMER UNTIL AGAR IS MELTED. ADD THE VEGETABLES, EXCEPT SCALLION, AND SIMMER JUST UNTIL THE PEAS ARE A BRIGHT GREEN. ADD THE SCALLIONS AND COOK ONE MINUTE LONGER. POUR INTO A MOLD OR BOWL. ALLOW IT TO SET AT ROOM TEMPERATURE. UNMOLD AND SERVE WITH A MAYONNAISE OR A VINEGAR DRESSING.

SEAFOOD ASPIC

SERVES 4

1 CUP CLEAR STOCK
1/2 CUP TOMATO JUICE
1 TSP. AGAR POWDER
1/2 TSP. WORCESTERSHIRE SAUCE
1 TSP. LEMON JUICE
PEPPER OR HOT SAUCE TO TASTE
1/4 TO 1/3 CUP COOKED SEAFOOD OR FISH
1/2 CUP VERY THINLY SLICED CELERY
1/2 RIPE AVOCADO, CUT IN CHUNKS

COMBINE STOCK, TOMATO JUICE, AND AGAR POWDER IN A SAUCEPAN AND HEAT TO BOILING, STIRRING UNTIL AGAR MELTS COMPLETELY. REMOVE FROM HEAT AND ADD THE WORCESTERSHIRE, LEMON, AND PEPPER. COOL UNTIL IT BEGINS TO JELL. STIR IN THE SEAFOOD, CELERY, AND AVOCADO. POUR INTO A MOLD AND REFRIGERATE UNTIL CHILLED. UNMOLD ON TO A PLATE AND GARNISH WITH PARSLEY. SERVE WITH A THINNED SOUR CREAM OR A MAYONNAISE DRESSING.

MINT JELLY

MAKES 1 PINT

2 CUPS APPLE JUICE
1 TEASPOON AGAR POWDER
  OR 1 TBSP. AGAR FLAKES
A BUNCH OF CLEAN, FRESH
  MINT OR 1 TO 4 TEASPOONS
  DRIED MINT

COMBINE THE APPLE JUICE AND AGAR IN A SAUCEPAN AND BRING TO A BOIL; REDUCE HEAT AND STIR UNTIL THE AGAR MELTS COMPLETELY. BRUISE THE MINT LEAVES AND, HOLDING THE BUNCH BY THE STEMS, SWISH THEM IN THE MIXTURE. TASTE UNTIL THE DESIRED INTENSITY OF MINT FLAVOR IS OBTAINED. ADD DRIED MINT (IF FRESH IS UNAVAILABLE) A LITTLE AT A TIME, STIRRING, WAITING FOR FLAVOR TO DEVELOP, AND ADJUSTING TO TASTE. POUR THE MIXTURE THROUGH A FINE SIEVE INTO JELLY GLASSES AND COOL COMPLETELY BEFORE SERVING. IF YOU WISH A GREEN COLOR, YOU SHOULD USE A GREEN FOOD COLORING.

STRAWBERRY-CANTALOUPE DESSERT

SERVES 4 TO 6

2 CUPS WATER
3 TEASPOONS AGAR POWDER
2 CUPS FRUIT JUICE
2 SMALL BOXES STRAWBERRIES
1 CANTALOUPE

COMBINE WATER AND AGAR IN A SAUCEPAN AND BRING TO A BOIL; REDUCE HEAT AND STIR UNTIL AGAR MELTS. MIX IN THE FRUIT JUICE. SET ASIDE UNTIL PARTIALLY SET.

WASH, HULL AND CUT THE STRAWBERRIES INTO PIECES. CLEAN SEEDS FROM THE CANTALOUPE AND SCOOP OUT IN BALLS OR CUT IN PIECES. FOLD THE FRUIT INTO THE JUICE MIXTURE. REFRIGERATE UNTIL CHILLED AND COMPLETELY FIRM. SERVE AS A DELIGHTFUL, LIGHT, SUMMER DESSERT.

COFFEE PARFAIT

SERVES 6

- 4 CUPS WATER
- 1/4 CUP HONEY OR MALT SYRUP
- 4 TSP. AGAR FLAKES OR 1 TSP. AGAR POWDER
- 2 CUPS COFFEE, POSTUM, OR PERO
- 1/8 TEASPOON SALT
- 1 CUP FIRM TOFU
- 2 TBSP. HONEY OR MALT SYRUP
- 1 TEASPOON ALMOND EXTRACT
- 1 TEASPOON VANILLA EXTRACT
- 1/8 TEASPOON SALT
- 1/2 CUP TOASTED, SLIVERED ALMONDS

BRING WATER, SWEETENER, AND AGAR TO A BOIL IN A SAUCEPAN, REDUCE HEAT AND STIR UNTIL AGAR MELTS. REMOVE FROM HEAT AND STIR IN THE COFFEE AND SALT. POUR ONE THIRD OF THIS MIXTURE INTO INDIVIDUAL SERVING GLASSES AND REFRIGERATE TO HASTEN SETTING.

COMBINE THE TOFU, SWEETENER, ALMOND EXTRACT, VANILLA, AND SALT IN A BLENDER AND BLEND UNTIL SMOOTH. AS SOON AS THE SURFACE OF THE AGAR MIXTURE BECOMES TACKY, GENTLY SPOON THE TOFU MIXTURE OVER IT. REFRIGERATE AGAIN UNTIL CHILLED. GARNISH WITH THE SLIVERED ALMONDS.

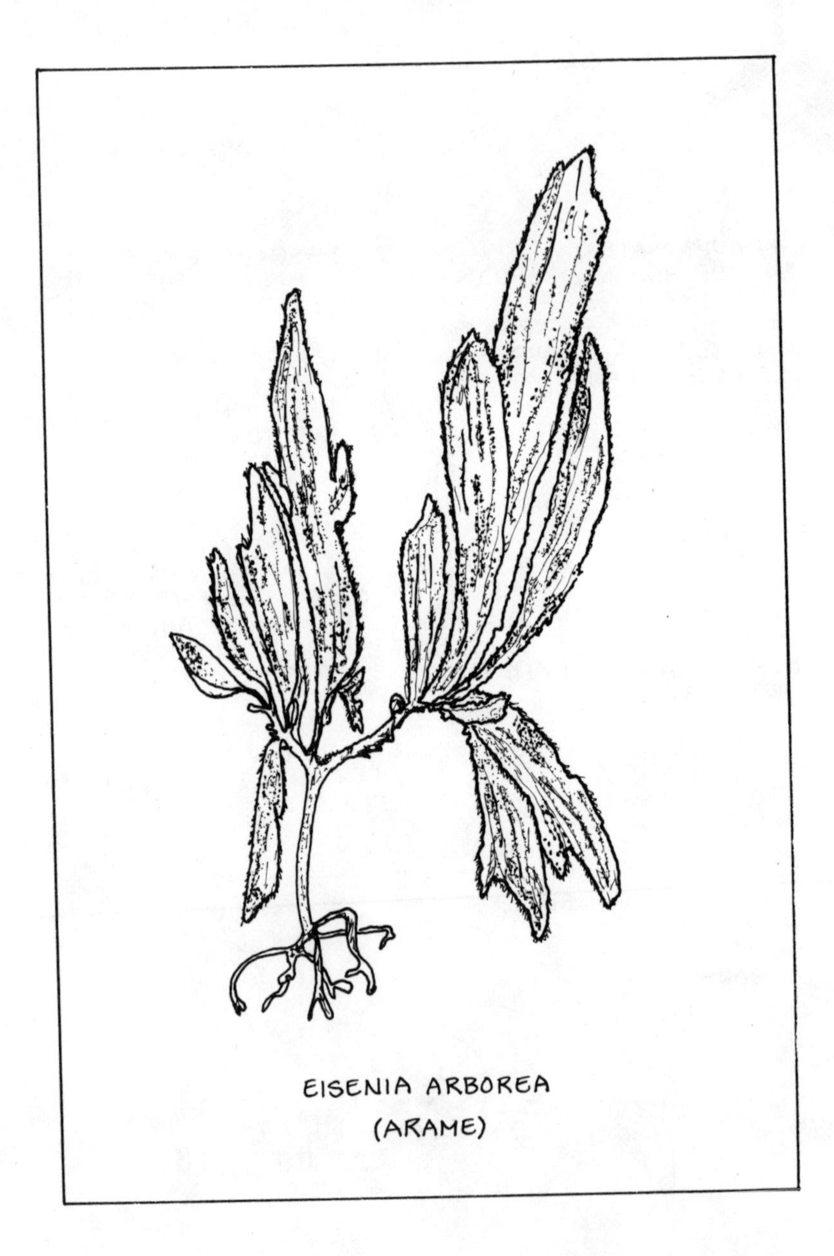

EISENIA ARBOREA
(ARAME)

ARAME

ARAME, MEANING "ROUGH MAIDEN", IS A DARK, YELLOWISH-BROWN, BROAD LEAFED PLANT THAT IS HARVESTED IN JAPAN AND CHINA AND GROWS ON THE PACIFIC COAST OF NORTH AND SOUTH AMERICA. IT CLOSELY RESEMBLES WAKAME AND KOMBU. THE 12" LONG BLADES ARE SUN-DRIED, RINSED, STEAMED OR PARBOILED, AND SUN-DRIED AGAIN. BLACK WHEN DRIED, ARAME IS SHREDDED INTO LONG THIN STRIPS RESEMBLING BLACK NOODLES.

HAVING A NUTTY FLAVOR AND A CRISP TEXTURE, ARAME IS SWEETER AND MORE DELICATE TASTING THAN HIJIKI, FOR WHICH IT IS OFTEN SUBSTITUTED. THE FLAVOR COMES FROM THE NATURAL SUGARS. ARAME IS RECOGNIZED AS AN EMERGENCY FOOD SUPPLY, MAINTAINING ITS TASTE, WHEN DRIED, FOR 2 TO 3 YEARS.

NUTRIENTS: ARAME IS AN EXCELLENT SOURCE OF CALCIUM AND PHOSPHORUS. IT CONTAINS IODINE AND ALGINIC ACID AND IS RICH IN IRON AND POTASSIUM.

PREPARATION: TO RECONSTITUTE, SIMPLY SOAK IN FRESH, COOL WATER FOR 15 MINUTES AND DRAIN. IT IS THEN READY FOR USE.

USES: ARAME MAY BE SUBSTITUTED FOR HIJIKI IN ANY RECIPE. IT IS EXCELLENT AS A SALAD VEGETABLE, AND IN SOUPS OR STEWS. IT CAN BE SAUTÉED AND STEAMED WITH LAND VEGETABLES AND USED AS A SIDE DISH TO COMPLEMENT GRAINS.

LAND AND SEA VEGETABLE SALAD WITH CARAWAY DRESSING

SERVES 4

SALAD

- 1/2 CUP DRIED ARAME
- 1 CUP ROMAINE LETTUCE, SHREDDED
- 1 CUP RED LEAF LETTUCE, SHREDDED
- 1/2 CUP BEETS, GRATED
- 1/2 CUP CARROTS, GRATED
- 1/2 CUP RED CABBAGE, THINLY SLICED
- 1/2 CUP SLICED MUSHROOMS
- 1/4 CUP SLICED RADISHES
- 1/4 CUP CAULIFLOWER FLOWERETS
- 20 TO 30 MOCHI CUBES, TOASTED

CARAWAY DRESSING

- 1 AVOCADO
- 1 FRESH TOMATO
- 1 LEMON, JUICED
- 1 TBSP. CARAWAY SEEDS
- 2 CLOVES GARLIC, MINCED
- 1 TBSP. POWDERED KELP (SEE KOMBU, PAGE 45)
- SALT AND PEPPER TO TASTE
- MORE TOMATO TO BLEND

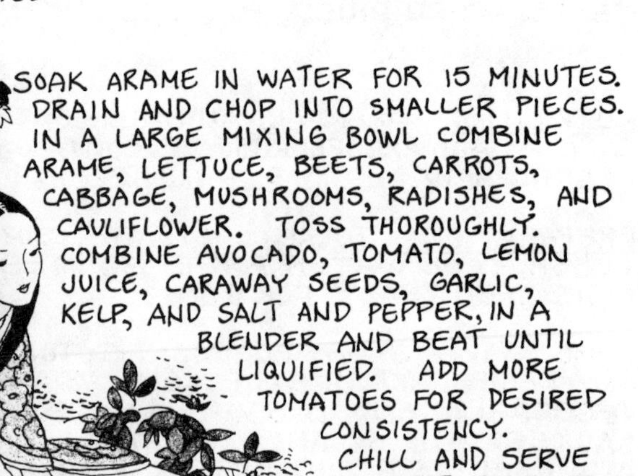

SOAK ARAME IN WATER FOR 15 MINUTES. DRAIN AND CHOP INTO SMALLER PIECES. IN A LARGE MIXING BOWL COMBINE ARAME, LETTUCE, BEETS, CARROTS, CABBAGE, MUSHROOMS, RADISHES, AND CAULIFLOWER. TOSS THOROUGHLY. COMBINE AVOCADO, TOMATO, LEMON JUICE, CARAWAY SEEDS, GARLIC, KELP, AND SALT AND PEPPER, IN A BLENDER AND BEAT UNTIL LIQUIFIED. ADD MORE TOMATOES FOR DESIRED CONSISTENCY. CHILL AND SERVE ON THE SIDE AS A DRESSING FOR THE SALAD. SPRINKLE SALAD WITH MOCHI CROUTONS AS A NUTRITIOUS GARNISH.

CREAMY TEMPEH SALAD WITH ARAME

SERVES 4

- 8 OZ. TEMPEH
- 1/2 CUP DRIED ARAME
- 2/3 CUP MAYONNAISE
- 1 CUP GREEN PEPPER, FINELY CHOPPED
- 1 CUP CELERY, FINELY CHOPPED
- 1/2 CUP CARROT, GRATED
- 1 CUP BROCCOLI, CUT INTO SMALL PIECES
- 3 GREEN ONIONS, CHOPPED
- 1/4 CUP SWEET RELISH
- 1 TSP. DILL WEED
- A DASH OF CAYENNE

CUT TEMPEH INTO SMALL CUBES. STEAM UNTIL TENDER. SOAK ARAME FOR 15 MINUTES. DRAIN AND CHOP FINE. COMBINE TEMPEH, ARAME, AND MAYONNAISE IN A LARGE BOWL MIXING THOROUGHLY. ADD GREEN PEPPER, CELERY, CARROT, BROCCOLI, GREEN ONION, RELISH, DILL WEED, AND CAYENNE, AND MIX LIGHTLY.

CHILL FOR 30 MINUTES. FOR A SPICY, HOT VARIATION, ADD HORSERADISH AND PREPARED MUSTARD TO TASTE.

TEMPEH CURRY WITH ARAME

SERVES 3 TO 4

16 OZ. TEMPEH
2 TBSP. OIL
1/2 CUP DRIED ARAME
1/2 CUP MINCED ONION
5 TBSP. BUTTER
6 TBSP. WHOLE WHEAT FLOUR
1 TBSP. CURRY POWDER
1 1/2 TEASPOON SALT
1/4 TEASPOON POWDERED GINGER
1 CUP VEGETABLE OR CHICKEN BROTH
2 CUPS MILK OR SOYMILK
1 TEASPOON LEMON JUICE
CHUTNEY, RAISINS, COCONUT, PEANUTS

CUT TEMPEH UP INTO SMALL 1/2" CUBES. SAUTÉ IN OIL FOR 15 MINUTES. SOAK ARAME IN WATER FOR 15 MINUTES. DRAIN AND CHOP FINELY. ADD TO TEMPEH AND STIR FRY FOR 5 MINUTES. SET ASIDE. SAUTÉ ONION IN BUTTER IN THE TOP OF A DOUBLE BOILER, OVER DIRECT HEAT, UNTIL TENDER. STIR IN FLOUR, CURRY, SALT, GINGER, BROTH, AND MILK. COOK OVER BOILING WATER IN DOUBLE BOILER, STIRRING CONSTANTLY UNTIL SMOOTH. ADD TEMPEH, ARAME, AND LEMON JUICE. SERVE WITH BROWN RICE OR À LA - WITH CHUTNEY, RAISINS, COCONUT, AND PEANUTS ON THE SIDE.

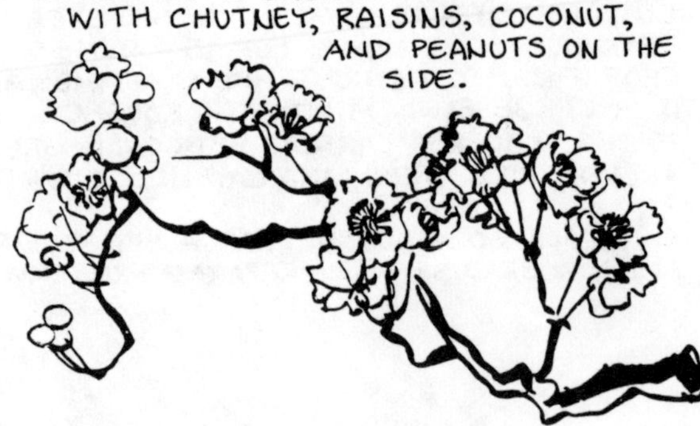

SCRAMBLED TOFU AND ARAME

SERVES 3-4

16 OZ. FIRM TOFU
3/4 CUP DRIED ARAME
1 ONION, CHOPPED FINELY
1 TBSP. OIL
1 MEDIUM BELL PEPPER, CHOPPED FINELY
4 TO 6 MUSHROOMS, CHOPPED
1 TBSP. MISO
1 CUP WARM WATER
TAMARI
1 TEASPOON DILL WEED
1 TEASPOON TURMERIC

DRAIN WATER FROM TOFU. WRAP TOFU BLOCK IN COTTON TOWEL AND PLACE A WEIGHT ON TOP TO EXTRACT MORE MOISTURE. SET ASIDE FOR 10 MINUTES. SOAK ARAME IN WATER FOR 15 MINUTES. DRAIN AND CHOP INTO SMALLER PIECES. MASH TOFU UP IN A SEPARATE BOWL WITH A FORK. THE MASHED TOFU WILL RESEMBLE WHITE SCRAMBLED EGGS. SAUTÉ ONION IN OIL FOR A MINUTE. ADD ARAME, BELL PEPPERS, AND MUSHROOMS. SAUTÉ FOR ANOTHER MINUTE AND ADD THE TOFU. ADD THE MISO DISSOLVED IN THE WARM WATER. SEASON WITH ADDITIONAL TAMARI TO TASTE. ADD DILL WEED AND TURMERIC. COOK ANOTHER 3 OR 4 MINUTES OVER LOW HEAT, STIRRING WELL. SERVE WITH WHOLE GRAIN TOAST, BAGELS, OR MILLET.

HIZIKIA FUSIFORME
(HIJIKI)

HIJIKI

HIJIKI IS A BLACKISH-BROWN, STRINGY PLANT HAVING A VERY DIFFERENT SHAPE AND TEXTURE FROM OTHER SEA GREENS. IT IS FOUND ON ROCKS IN THE LOW TIDAL AREAS OFF THE COAST OF JAPAN AND CHINA. IT IS A MAJOR FOOD IN THE ORIENT.

FOLK LEGEND REGARDS HIJIKI AS AN EXCELLENT FOOD FOR MAINTAINING HEALTHY HAIR; JAPANESE GIRLS WERE TOLD TO CONSUME BOWLFULS OF THE SHINY BLACK FLOWER PETALS.

HIJIKI IS HARVESTED IN THE SPRING. THE PLANTS ARE STEAMED OR PARBOILED TO ELIMINATE UNPLEASANT ASTRINGENCY AND SUN DRIED. PACKAGED HIJIKI RESEMBLES A TANGLE OF BLACK SHOELACES. IT IS MUCH MILDER THAN DULSE, KOMBU, OR NORI, HAVING A CRUNCHY TEXTURE AND A DELICIOUS NUT-LIKE FLAVOR.

NUTRIENTS: HIJIKI HAS MORE CALCIUM THAN ANY OTHER FOOD SOURCE, 1400 MG. FOR EACH 100 GRAMS OF DRIED PLANT. IT ALSO CONTAINS LARGE QUANTITIES OF VITAMINS A, B_1, B_2, PHOSPHORUS, AND IRON.

PREPARATION: HIJIKI QUADRUPLES IN VOLUME, SO BE CAREFUL NOT TO START WITH TOO MUCH. WASH AND SOAK IN WARM WATER TO EXPAND. AFTER 5 MINUTES, CAREFULLY LIFT IT OUT OF THE SOAKING PAN, THIS WILL HELP KEEP ANY SAND ON THE BOTTOM OF THE PAN. DRAIN AND RINSE WELL, PRESSING OUT EXCESS WATER. DO NOT SOAK OVER LONG AS IT WILL GET SOGGY. USE MEDIUM HEAT FOR SAUTÉING.

USES: HIJIKI CAN BE SUBSTITUTED FOR ARAME IN ANY RECIPE. IT IS GOOD SAUTÉED WITH OTHER VEGETABLES OR CRUMBLED INTO SOUPS OR SALADS.

TOFU HIJIKI SALAD

SERVES 2 TO 4

SALAD

- 1/4 CUP DRIED HIJIKI
- 8 OZ. FIRM TOFU
- 4 GREEN ONIONS SLICED VERY THIN
- 2 STALKS CELERY, CHOPPED
- 2 CARROTS, GRATED
- 1/4 BUNCH PARSLEY, CHOPPED
- 3 TBSP. NUTRITIONAL YEAST
- 1/2 TEASPOON SALT (OPTIONAL)
- 2 TBSP. SESAME SEEDS (TOASTED)

DRESSING

- 1/4 CUP SALAD OIL
- 1 TBSP. LEMON JUICE
- 2 CLOVES GARLIC, MINCED
- 2 TBSP. TAMARI
- 1 TEASPOON DILL WEED
- 1/2 TEASPOON PAPRIKA

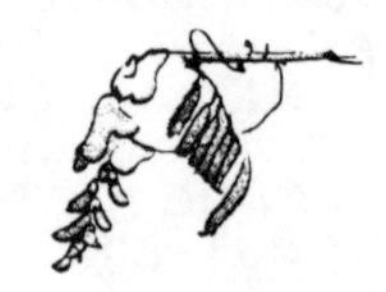

SOAK HIJIKI IN WARM WATER FOR 5 MINUTES, DRAIN, CUT INTO SMALL PIECES, AND DRY. DRAIN TOFU. WRAP A CLOTH TOWEL AROUND IT AND PLACE A WEIGHT ON IT FOR 10 MINUTES TO EXTRACT MORE MOISTURE. CRUMBLE TOFU ALONG WITH HIJIKI INTO A MIXING BOWL AND MASH WITH A POTATO MASHER. ADD GREEN ONIONS, CELERY, CARROTS, PARSLEY, NUTRITIONAL YEAST, SALT, AND TOASTED SESAME SEEDS. MIX WELL. IN A BLENDER, LIQUIFY OIL, LEMON JUICE, GARLIC, TAMARI, DILL WEED, AND PAPRIKA. POUR THIS DRESSING OVER TOFU MIXTURE AND TOSS LIGHTLY. SERVE AS A TOPPING FOR A GREEN SALAD WITH CRACKERS OR CHIPS; IN PITA BREAD WITH AVOCADO AND SPROUTS AS A SANDWICH, OR WITH A LIGHT GRAIN.

TOFU CASHEW CURRY WITH HIJIKI

SERVES 2-3

- 1/2 CUP DRIED HIJIKI
- 3/4 CUP RAW CASHEWS
- 10 OZ. FIRM TOFU
- 3 TBSP. SAFFLOWER OIL
- 2 MEDIUM ONIONS, SLICED IN WEDGES
- 2 CLOVES GARLIC, MINCED
- 1/2 TEASPOON KELP (SEE KOMBU, PAGE 45)
- 1/2 TEASPOON GINGER
- 1 TBSP. TAMARI
- 1 TBSP. CURRY POWDER (OR TO TASTE)
- 2 MEDIUM GREEN PEPPERS, SLICED VERY THINLY
- 1 CUP GREEN PEAS (FRESH OR FROZEN)

SOAK HIJIKI IN WATER FOR 5 MINUTES. DRAIN AND DRY ON PAPER TOWELS. CUT INTO SMALLER PIECES. ROAST CASHEWS ON A COOKIE SHEET OR BAKING PAN IN A 300° OVEN FOR 20 MINUTES (STIRRING OCCASIONALLY TO PREVENT BURNING) UNTIL THEY JUST START TO TURN LIGHT BROWN. SET ASIDE. DRAIN AND PRESS TOFU BLOCK IN A TOWEL WITH SOME WEIGHT ON THE TOP. CUT INTO 1" CUBES. HEAT OIL IN A WOK OR LARGE SKILLET. STIR-FRY ONIONS AND GARLIC FOR 2 MINUTES. ADD HIJIKI, TOFU, KELP, GINGER, TAMARI, CURRY, AND STIR-FRY FOR A FEW MORE MINUTES. ADD GREEN PEPPERS AND GREEN PEAS AND STIR-FRY AGAIN. CONTINUE COOKING ON LOW HEAT, STIRRING OFTEN. ADD A SMALL AMOUNT OF WATER IF NEEDED. COOK FOR 3 TO 5 MORE MINUTES UNTIL VEGETABLES ARE COOKED AND FLAVORS ARE MELDED. REMOVE FROM HEAT AND LIGHTLY TOSS IN THE ROASTED CASHEWS. SERVE IMMEDIATELY WITH STEAMED RICE OR ÁLÀ.

PEA AND SEA SALAD

SERVES 3 TO 4

1/4 CUP DRIED HIJIKI
1 CUP FRESH PEAS
1 CUP CHOPPED COMFREY OR SHREDDED CABBAGE
1/2 CUP GRATED CARROTS
2 TBSP. SCALLIONS, FINELY CHOPPED
SALAD DRESSING

SOAK HIJIKI IN WARM WATER FOR 5 MINUTES. DRAIN AND CUT INTO SMALLER PIECES. COMBINE VEGETABLES IN A SALAD BOWL AND TOSS WITH YOUR FAVORITE SALAD DRESSING.

HIJIKI-GARBANZO SALAD

SERVES 4

1 1/2 CUPS DRY GARBANZO BEANS (CHICKPEAS)
1/2 TEASPOON SALT
1/2 CUP DRIED HIJIKI
6 TBSP. TAHINI
6 TBSP. WATER
3 TBSP. LEMON JUICE
1 1/2 TEASPOONS POWDERED KELP (SEE KOMBU PAGE 45)
CAYENNE AND SALT TO TASTE

SOAK GARBANZOS OVERNIGHT; DRAIN AND THEN BOIL IN SALTED WATER FOR ABOUT 1 HOUR, OR UNTIL TENDER. RINSE, DRAIN AND COOL. SOAK HIJIKI FOR 5 MINUTES; DRAIN, CHOP AND MIX WITH GARBANZOS IN A LARGE BOWL. COMBINE TAHINI, WATER, LEMON JUICE, KELP, CAYENNE AND SALT IN A BLENDER AND LIQUIFY. POUR OVER GARBANZOS AND HIJIKI, TOSS, AND SERVE.

CARROT HIJIKI STIR-FRY

SERVES 4

1/2 CUP DRIED HIJIKI
1 TBSP. OIL
3/4 CUP EACH ONIONS, CARROTS, GREEN BEANS, SLICED
2 TBSP. TAMARI
1 TBSP. HONEY
1 TEASPOON FRESH GRATED GINGER
A PINCH OF CAYENNE
20 TO 30 MOCHI CUBES, TOASTED

SOAK HIJIKI FOR 5 MINUTES AND DRAIN. CUT INTO SMALL PIECES. STIR FRY ONIONS IN OIL UNTIL TRANSLUCENT; ADD CARROTS AND GREEN BEANS AND STIR FRY 2 MINUTES. ADD HIJIKI AND TOSS OVER HEAT FOR 2 MINUTES. ADD TAMARI, HONEY, GINGER, AND CAYENNE. COVER AND SIMMER FOR 10 MINUTES. REMOVE COVER AND STEAM AWAY EXCESS LIQUID. REMOVE FROM HEAT TO A SERVING BOWL AND SPRINKLE WITH TOASTED MOCHI CUBES. SERVE IMMEDIATELY WITH RICE.

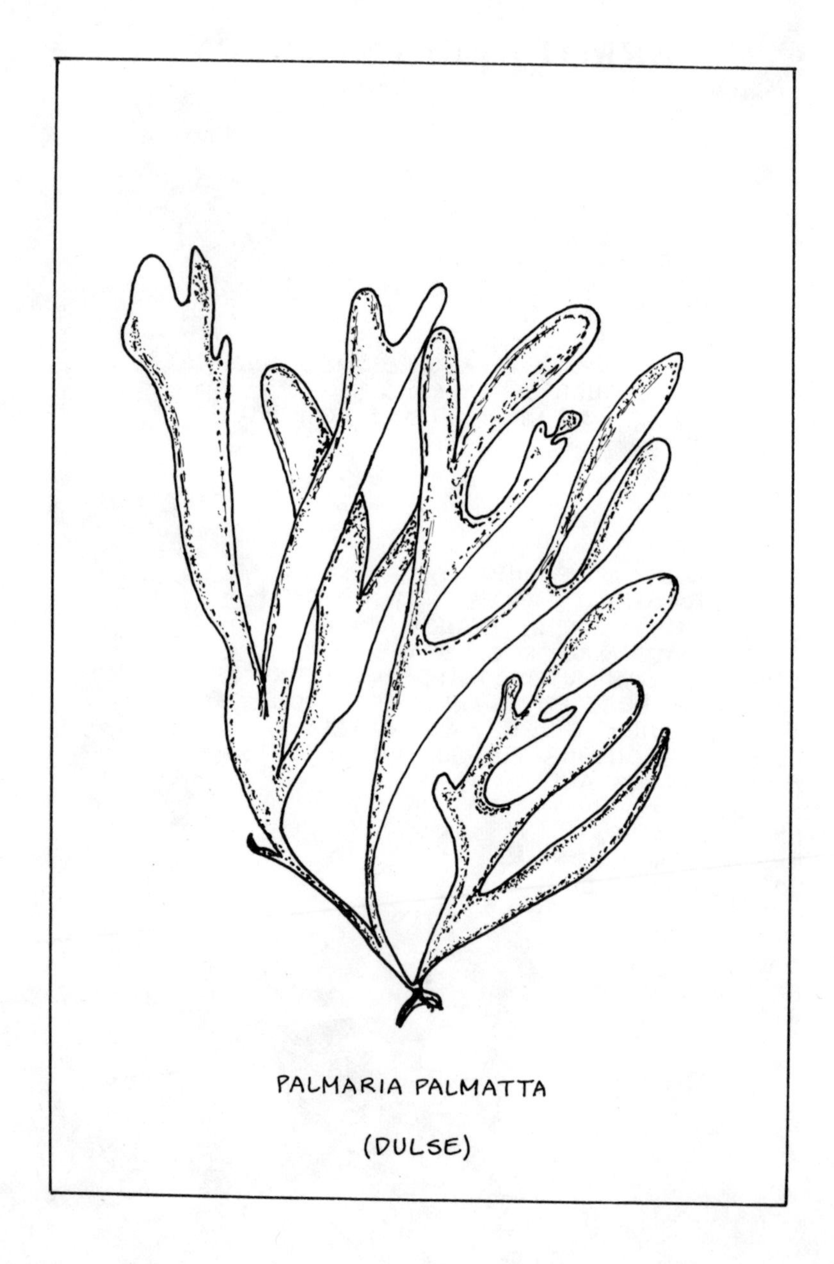

PALMARIA PALMATTA

(DULSE)

DULSE

DULSE IS OFTEN CALLED "RED KALE" OR THE "BEEF JERKY OF THE SEA." IT HAS A CHEWY, STRINGY TEXTURE AND A DELICIOUS, NUT-LIKE FLAVOR. DULSE IS WIDELY FOUND IN THE COLDER WATERS OF BOTH THE ATLANTIC AND PACIFIC OCEANS. IT IS GATHERED COMMERCIALLY FROM THE BAY OF FUNDY IN CANADA AND FROM THE COASTS OF WASHINGTON AND CALIFORNIA.

IN THE EARLY 1900'S IT WAS COMMON TO SEE TANGY, DRIED DULSE IN NEW ENGLAND RAILROAD STATIONS. IT WAS SERVED IN BARS AS A SNACK, ITS SALTINESS INCREASING THIRST AND REVENUE. THE USE OF DULSE DATES BACK TO THE 8TH CENTURY IN THE BRITISH ISLES, WHEN IT WAS COMMONLY EATEN WITH DRIED FISH, POTATOES, AND BUTTER.

DULSE IS A DEEP, REDDISH PURPLE COLOR; ITS FLAT, FAN-SHAPED BLADES ARE UP TO 16" LONG. THEY ARE HARVESTED FROM ROCKS AT LOW TIDE DURING THE SUMMER. THE LEAVES ARE RINSED, COMPRESSED, AND DRIED. IT IS PACKAGED READY TO EAT.

<u>NUTRIENTS:</u> DULSE HAS THE HIGHEST CONCENTRATION OF IRON OF ANY FOOD. IT IS VERY RICH IN POTASSIUM AND MAGNESIUM, AND A GOOD SOURCE OF CALCIUM, IODINE, AND PHOSPHORUS. IT CONTAINS VITAMINS A, B_2, B_6, C, E, AND MANY TRACE MINERALS.

<u>PREPARATION:</u> DULSE DOES NOT REQUIRE PRE-SOAKING. RINSE BRIEFLY UNDER TAP WATER TO REMOVE SAND AND IMPURITIES.

<u>USES:</u> CAN BE EATEN AS IT COMES AS A SNACK. MAY BE ADDED TO RAW VEGETABLE SALAD, OR ADDED DRY AND CRUMBLED TO SOUPS, STEWS, RELISHES, BREADS, AND FRITTERS. IT CAN BE DRY ROASTED AND GROUND TO MAKE A CONDIMENT AND DEEP FRIED AS A SNACK.

DULSE SALAD WITH TAHINI DRESSING

SERVES 8 TO 10

SALAD

SALAD OF YOUR CHOICE
1 CUP DRIED DULSE

DRESSING

2 CUPS OIL
1 CUP TAHINI
1/3 CUP TAMARI, OR TO TASTE
1/2 CUP LEMON JUICE
3/4 CUP WATER
2 STALKS CELERY, DICED
1/2 ONION, CHOPPED
1/2 BELL PEPPER, CHOPPED FINELY

MAKE YOUR FAVORITE FRESH SALAD LARGE ENOUGH FOR 8 TO 10 PEOPLE. RINSE DULSE AND TEAR INTO SMALL PIECES. TOSS INTO SALAD. BLEND THE OIL, TAHINI, TAMARI, LEMON JUICE, WATER, CELERY, ONION, AND BELL PEPPER IN A BLENDER UNTIL THOROUGHLY MIXED. REFRIGERATE. SERVE DRESSING ON THE SIDE. THE DULSE ADDS A WONDERFUL FLAVOR AS WELL AS ADDITIONAL NUTRITION TO YOUR SALAD.

DULSE CHIPS

4 OZ. DRIED DULSE

PLACE DRIED SHREDDED DULSE ON A COOKIE SHEET AND BAKE 5 TO 10 MINUTES IN 350° OVEN OR UNTIL IT CRUSHES EASILY. SERVE AS WHOLE PIECES (SEA CHIPS) OR PULVERIZE FOR A HOMEMADE SEA GREEN POWDER. DO NOT ALLOW CHIPS TO TURN BLACK OR THE TASTE WILL BE BITTER. SPRINKLE ON GRAINS OR VEGETABLES.

LENTIL BARLEY SOUP WITH DULSE

SERVES 4 TO 6

- 1/3 CUP BARLEY
- 2/3 CUP DRY LENTILS
- 1 TEASPOON SALT
- 1 1/2 QUARTS WATER
- 1 TBSP. OIL
- 2 MEDIUM ONIONS, CUT INTO CRESCENTS
- 1 STALK CELERY, SLICED
- 1 CARROT SLICED
- 1/4 CUP CHOPPED PARSLEY
- 1/4 TEASPOON THYME
- 1/4 TEASPOON OREGANO
- 1 TEASPOON TAMARI
- 1/2 CUP DRIED DULSE

WASH THE BARLEY AND LENTILS AND SOAK OVERNIGHT IN 3 CUPS OF WATER. IN THE MORNING, ADD 3 MORE CUPS OF WATER, AND SALT, AND COOK UNTIL THE GRAINS ARE SOFT, ABOUT 1 1/2 HOURS.

HEAT THE OIL IN A SKILLET AND SAUTÉ THE ONION, CELERY, CARROT, AND PARSLEY UNTIL TENDER. ADD THYME AND OREGANO AND ADD TO THE SOUP POT. COOK FOR 45 MINUTES. TASTE AND SEASON WITH TAMARI. ADD ADDITIONAL WATER, IF NECESSARY.

REMOVE FROM HEAT AND STIR IN SHREDDED DULSE THAT HAS BEEN RINSED AND CUT INTO PIECES. SERVE WITH WHOLE GRAIN BREAD AND A GREEN SALAD.

CREAM OF DULSE AND POTATO SOUP

SERVES 4 TO 6

1/2 CUP DRIED DULSE
1 1/2 TBSP. OIL
1 1/2 LBS. POTATOES, CUBED
1 LEEK, CHOPPED (WHITE PORTION ONLY)
1 1/2 TEASPOONS ROSEMARY
1 1/2 TEASPOONS THYME
6 CUPS SOUP STOCK (ANY TYPE)
2 TBSP. RICE FLOUR OR CORNSTARCH
1/4 CUP COLD WATER
1/2 CUP MINCED SCALLIONS

SOAK DULSE BRIEFLY IN WATER TO REMOVE DEBRIS AND EXCESS SALT. SHRED INTO SMALLER PIECES AND SET ASIDE TO DRAIN. HEAT OIL IN A SOUP POT. SAUTÉ POTATOES AND LEEK UNTIL WELL COATED WITH OIL. ADD ROSEMARY AND THYME. ADD SOUP STOCK, COVER AND SIMMER UNTIL POTATOES ARE TENDER. LET SOUP COOL AND THEN PURÉE THE MIXTURE WITH DULSE IN A BLENDER. RETURN PURÉE TO POT. THICKEN BY ADDING SOME RICE FLOUR OR CORNSTARCH MIXED WITH A LITTLE COLD WATER. HEAT AND SIMMER FOR 5 MINUTES. SERVE GARNISHED WITH SCALLIONS.

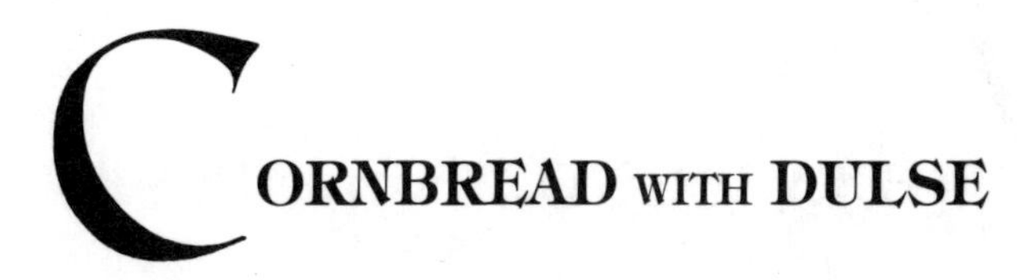

CORNBREAD WITH DULSE

PRE-HEAT OVEN TO 350°

SERVES 4 TO 6

- 1 CUP CORN MEAL
- 1 CUP WHOLE WHEAT FLOUR
- 3 TBSP. 7 GRAIN CEREAL (DRY)
- 2 TSP. BAKING POWDER
- 1/2 TSP. BAKING SODA
- 1/2 TSP. SALT
- 2 TBSP. NUTRITIONAL YEAST
- 1 TSP. CUMIN
- 1/3 CUP DRIED DULSE SHREDDED INTO SMALL PIECES
- 1 EGG
- 3-4 CUPS BUTTERMILK
- 1/4 CUP OIL
- 1/4 CUP DARK MOLASSES
- 1 CARROT, GRATED
- 1 BELL PEPPER, CHOPPED FINELY
- 1 FRESH COB OF CORN, SHAVED (OPTIONAL)

IN A LARGE MIXING BOWL COMBINE CORN MEAL, FLOUR, CEREAL, BAKING POWDER, BAKING SODA, SALT, NUTRITIONAL YEAST, CUMIN, AND DRIED DULSE. IN A SEPARATE BOWL BEAT THE EGG AND ADD 3 CUPS OF BUTTERMILK, OIL, AND MOLASSES. MIX THOROUGHLY AND ADD TO THE DRY INGREDIENTS. MIX ONLY UNTIL MOISTENED. FOLD IN CARROT, BELL PEPPER, AND CORN QUICKLY. CONSISTENCY SHOULD BE SIMILAR TO A THICK SMOOTHIE OR MILK SHAKE. ADD MORE BUTTERMILK IF TOO DRY.

POUR INTO A GREASED BAKING PAN AND BAKE FOR 40 TO 50 MINUTES AT 350°.

VEGETABLE STEW WITH DULSE

SERVES 4

- 1/2 CUP OIL
- 1 LARGE ONION, CHOPPED
- 2 CLOVES GARLIC, MINCED
- 16 OZ. FIRM TOFU, CUT INTO 1" CUBES
- 2 MEDIUM CARROTS, SLICED THINLY
- 1 HEAD CABBAGE, CHOPPED
- 1 GREEN PEPPER, CHOPPED
- 1 ZUCCHINI, CHOPPED
- 2 STALKS CELERY, CHOPPED
- 1 CUP WATER
- 1 BAY LEAF
- A PINCH OF CAYENNE
- 1/4 CUP TAMARI
- 1/2 CUP DRIED DULSE, SHREDDED
- 2 TBSP. WHOLE WHEAT PASTRY FLOUR
- 1/4 CUP COLD WATER

HEAT OIL IN A GOOD SIZED POT. SAUTÉ ONION AND GARLIC UNTIL ALMOST TRANSLUCENT. AFTER TOFU HAS BEEN DRAINED, PRESSED AND CUT INTO CUBES, ADD TO POT AND SAUTÉ A BIT MORE. ADD CARROTS, CABBAGE, GREEN PEPPER, ZUCCHINI, CELERY, WATER, BAY LEAF, CAYENNE AND BRING TO A BOIL. REDUCE HEAT AND SIMMER COVERED FOR 30 MINUTES. ADD DRIED DULSE AND COOK 10 MINUTES MORE. MIX FLOUR WITH WATER, ADD TO STEW, AND COOK UNTIL THICKENED.

SERVE WITH BROWN RICE OR WHEAT PILAF.

SEA GREEN QUICHE

SERVES 6

PRE HEAT OVEN TO 350°

PREPARE A SINGLE CRUST OF YOUR CHOICE FOR A 9" PIE. PRE-BAKE FOR 5 MINUTES. THEN MAKE FILLING;

1 TBSP. SESAME OIL
2 CLOVES GARLIC, MINCED
1 MEDIUM ONION, CHOPPED
1 1/2 CUPS BROCCOLI FLOWERS
1/4 CUP FRESHLY CHOPPED PARSLEY
1/4 CUP DRIED HIJIKI,
1 TBSP. POWDERED KELP (SEE KOMBU PAGE 45)
1 TBSP. POWDERED DULSE, (SEE DULSE PAGE 34)
1/4 TEASPOON CUMIN
1/4 TEASPOON BASIL
1/4 TEASPOON SALT
1/4 TEASPOON THYME
A DASH OF CAYENNE
1/2 CUP CHOPPED SPINACH
1/2 CUP TAHINI
1 TBSP. HONEY
16 OZ. FIRM TOFU, (DRAINED AND MASHED)
1 CUP WATER
1/2 TEASPOON SALT (TO TASTE)
3 TBSP. ARROWROOT POWDER OR CORNSTARCH

SOAK HIJIKI IN WATER FOR 5 MINUTES. DRAIN, DRY, AND CUT INTO SMALL PIECES. HEAT OIL IN A SKILLET OR WOK. ADD GARLIC, ONION, BROCCOLI, PARSLEY, HIJIKI, KELP, DULSE, CUMIN, BASIL, SALT, THYME, AND CAYENNE AND SAUTÉ FOR 5 MINUTES. PLACE THIS MIXTURE IN A PARTIALLY COOKED PIE SHELL. LAYER SPINACH NEXT ON TOP OF SAUTÉED MIXED VEGETABLES. IN A BLENDER, COMBINE TAHINI, HONEY, TOFU, WATER, SALT (OPTIONAL), AND ARROWROOT. MIX THOROUGHLY TO GET A SMOOTH, THICK TEXTURE. POUR THIS MIXTURE OVER THE VEGETABLES. BAKE AT 350° FOR 30 MINUTES AND SERVE WITH A GREEN SALAD.

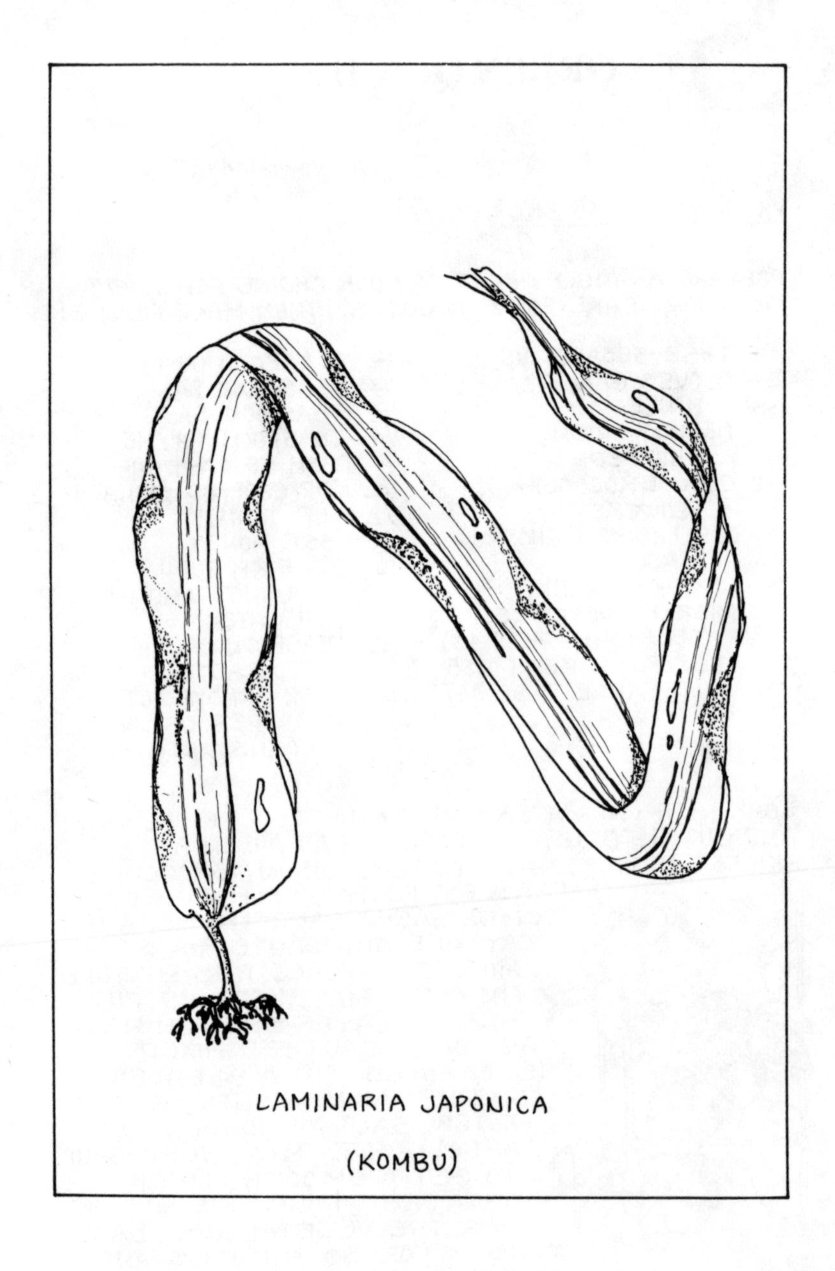

LAMINARIA JAPONICA

(KOMBU)

KOMBU

KOMBU IS THE SEAWEED _LAMINARIA_ AND VARIETIES OF IT ARE FOUND ON ALMOST EVERY COAST. IT VARIES IN LENGTH FROM 2 TO 24 FEET. THE FRONDS ARE HARVESTED IN SUMMER. BLADES OF MORE THAN ONE YEAR'S GROWTH ARE PREFERRED. IT IS GATHERED WITH LONG POLES WITH HOOKS OR FORKS ON THE ENDS, AND DRIED IN THE SUN.

KOMBU IS A ROBUSTLY FLAVORED SEA GREEN THAT HAS BEEN EATEN BY PEOPLE OF WALES, RUSSIA, IRELAND, AND JAPAN FOR HUNDREDS OF YEARS. IN RECENT TIMES, THE EATING OF THIS SEAWEED, COMMONLY CALLED "KELP," HAS GONE OUT OF FASHION IN THE WEST AND IT IS USED PRIMARILY AS A SOURCE OF ALGIN.

KOMBU HAS A GREENISH-BLACK COLOR. IT IS SLIGHTLY SWEET TO THE TASTE. IT CONTAINS GLUTAMIC ACID, WHICH IS RESPONSIBLE FOR ITS FLAVOR. MONOSODIUM GLUTAMATE, OR MSG, MADE FROM KOMBU.

NUTRIENTS: KOMBU IS A GOOD SOURCE OF POTASSIUM, SODIUM, AND VITAMINS A AND B.

PREPARATION: TO CLEAN KOMBU, OPEN THE FOLDED PIECE AND BRUSH OFF ANY SAND WITH YOUR FINGERS OR BY SLAPPING EACH PIECE WITH YOUR HANDS. WASH THOROUGHLY UNDER RUNNING WATER.

USES: KOMBU IS USED MOST IMPORTANTLY AS A SOUP STOCK, ALSO CALLED "KOMBU-DASHI." IT CAN ALSO BE DEEP FRIED OR BAKED. IF ADDED TO SOUPS OR STEWS IT MUST BE REMOVED BEFORE SERVING. IT HAS BEEN KNOWN TO BE USED AS A CHEWING TOBACCO SUBSTITUTE.

KOMBU SOUP STOCK

2 STRIPS KOMBU

KOMBU SOUP STOCK, CALLED "DASHI" IS THE BASIS FOR MISO SOUPS, CLEAR SOUPS, NOODLE (SOBA) BROTH, AND BOILED VEGETABLES.

CLEAN AND WASH KOMBU TO REMOVE ALL SAND AND GRIT. SOAK KOMBU IN 6 CUPS OF WATER OVERNIGHT. THEN HEAT UNTIL THE WATER JUST BEGINS TO BOIL. REMOVE THE KOMBU. DO NOT BOIL THE STOCK WITH THE KOMBU AS IT WILL BECOME BITTER AND STICKY.

BEANS AND KOMBU

SOAK ANY DRY BEANS OVERNIGHT. DISCARD SOAKING WATER, COVER BEANS WITH FRESH WATER, ADD SOME SALT, AND COOK, UNCOVERED, FOR 30 MINUTES. ADD 2 OR 3 STRIPS OF KOMBU. COVER POT AND COOK UNTIL BEANS ARE SOFT. THIS IS EXCELLENT FOR BLACK, ADUKI, CHICKPEAS, KIDNEY, PINTO, OR ANY OTHER BEAN.

KOMBU ADDED TO BEANS HELPS THE BREAK DOWN OF PROTEINS IN DIGESTION AND SOFTENS BEANS QUICKLY. IT IS REPORTED THAT USING KOMBU IN THIS MANNER REDUCES FLATULENCE CAUSED BY EATING BEANS.

RED SNAPPER AND TOFU

SERVES 4

1 STRIP KOMBU, ABOUT 6 INCHES LONG
1/2 CAKE FIRM TOFU
1 LB. RED SNAPPER FILET
1 LEMON RIND (OR TO TASTE) CUT INTO 1/8" STRIPS
3 TBSP. TAMARI
JUICE OF 1 LEMON
1 TSP. HONEY

PLACE KOMBU IN THE BOTTOM OF A STEAMER POT. COVER WITH WATER AND SOAK WHILE PREPARING OTHER INGREDIENTS. CUT TOFU INTO 1" CUBES. CUBE FISH. PLACE TOFU AND FISH IN STEAMER BASKET WITH LEMON RIND STRIPS. COVER AND STEAM FOR 15 MINUTES. IN A SMALL SAUCEPAN MIX TAMARI, LEMON JUICE, AND HONEY AND HEAT ONLY UNTIL HONEY MELTS. ARRANGE THE STEAMED FISH, TOFU, AND LEMON PEEL ON A PLATTER. POUR THE TAMARI MIXTURE OVER AND SERVE VERY HOT WITH STEAMED WHITE OR BROWN RICE AND A GREEN VEGETABLE. DISCARD THE KOMBU.

KOMBU TEMPURA

SERVES 3 TO 4

16 ~ 1/2" × 4" STRIPS OF DRIED KOMBU
1/3 CUP WHOLE WHEAT FLOUR
1 TBSP. CORNSTARCH
2 TEASPOONS SESAME SEEDS
1 EGG, BEATEN
3 TEASPOONS FRESH FINELY CHOPPED PARSLEY
A DASH OF CAYENNE
~ WATER TO THIN
2 CUPS PEANUT OIL
1 CUP TOASTED SESAME OIL
1 CUP SAFFLOWER OIL

CLEAN AND CUT KOMBU WITH A SCISSORS INTO APPROPRIATE SIZED PIECES AND SET ASIDE. COMBINE FLOUR, CORNSTARCH, SESAME SEEDS, EGG, PARSLEY, AND CAYENNE IN A BOWL AND MIX THOROUGHLY. ADD WATER TO THIN TO CONSISTENCY OF HEAVY CREAM.
HEAT THE DIFFERENT OILS TOGETHER IN A LARGE, DEEP SKILLET OR WOK TO 345.° TEST OIL BY DROPPING A FEW DROPS OF BATTER INTO THE OIL. IF THE OIL IS THE RIGHT TEMPERATURE, THE BATTER SHOULD BUBBLE RAPIDLY AND FLOAT TO THE SURFACE WITHIN 10 TO 15 SECONDS. DIP KOMBU IN BATTER AND PLACE IN HOT OIL. FRY BOTH SIDES UNTIL BATTER IS GOLDEN BROWN AND CRISP TO THE TOUCH OF A SLOTTED SPOON. REMOVE WITH A SLOTTED SPOON, ALLOWING EXCESS OIL TO DRIP BACK INTO THE PAN. PLACE DEEP FRIED KOMBU ON PAPER TOWEL TO ABSORB OIL. SERVE WITH TAMARI FOR DIPPING. INCREASE BATTER INGREDIENTS AND CUT UP CARROTS, BROCCOLI, ZUCCHINI, AND MUSHROOMS, IF YOU LIKE, AND DEEP FRY. DEEP FRIED KOMBU TASTES VERY MUCH LIKE BACON.

KOMBU OR KELP POWDER

4 1/2" × 8" STRIPS OF DRIED KOMBU
NUTRITIONAL YEAST FLAKES

WASH KOMBU WELL AND PAT DRY. BAKE IN A 350° OVEN FOR 10 TO 15 MINUTES UNTIL IT IS CRISP AND GREEN IN COLOR. COOL ON A RACK AND PULVERIZE IN A BLENDER OR SURIBACHI. MIX WITH YEAST FLAKES, IF DESIRED. USE AS A SEASONING OR IN PLACE OF TABLE SALT.

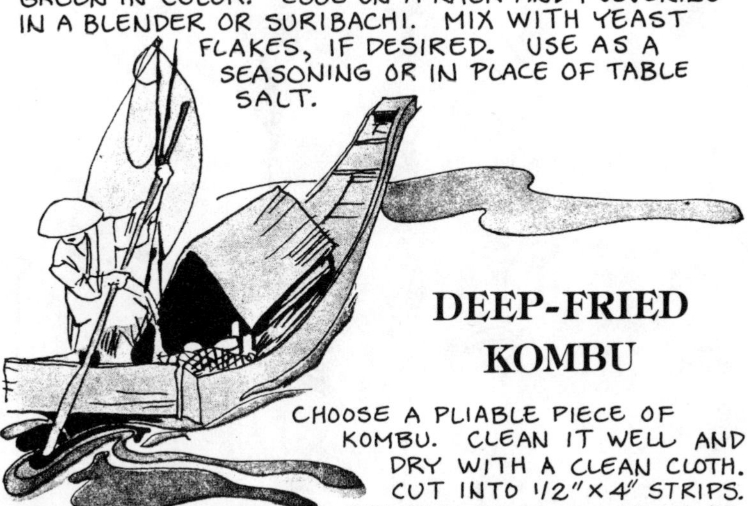

DEEP-FRIED KOMBU

CHOOSE A PLIABLE PIECE OF KOMBU. CLEAN IT WELL AND DRY WITH A CLEAN CLOTH. CUT INTO 1/2" × 4" STRIPS. TIE EACH PIECE IN A KNOT AT THE CENTER AND DEEP FRY IN HOT PEANUT OIL UNTIL CRISP. KOMBU DEEP-FRIES QUICKLY, SO BE CAREFUL NOT TO BURN IT. IF YOU WISH TO SOAK THE KOMBU IN WATER FOR 2 1/2 HOURS, DRAIN, THEN FREEZE THE STRIPS, THEY WILL NOT BE SO TOUGH. SOAK FROZEN KOMBU IN TAMARI OR OTHER MARINADE UNTIL THAWED, DRAIN EXCESS LIQUID AND THEN DEEP-FRY. SERVE AS A GARNISH FOR VEGETABLES AND GRAINS OR AS A SNACK WITH BEER.

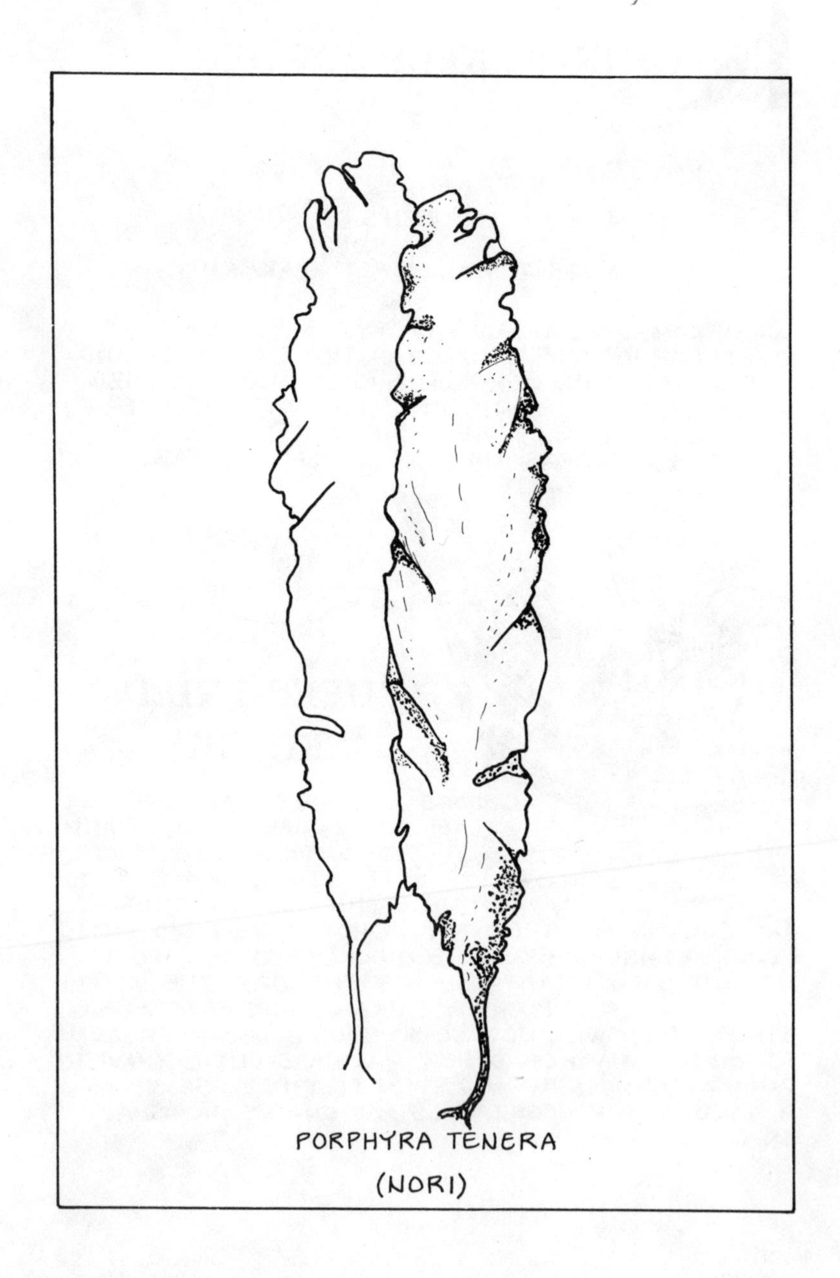

PORPHYRA TENERA

(NORI)

NORI

NORI IS A RED ALGAE OF THE WIDELY DISTRIBUTED PORPHYRA SPECIES. IT IS A BRIGHT LAVENDER IN THE WATER. THE TEN INCH FLAT BLADES GROW LIKE SEMI-TRANSPARENT, RUFFLED FANS, UNDULATING IN THE INTERTIDAL ZONE. NORI DRIES TO A DARK PURPLE OR BLACK COLOR THAT TURNS IRIDESCENT WHEN TOASTED.

THE MOST POPULAR NORI IN JAPAN IS PROPHYRA TENERA, BUT IT IS NO LONGER HARVESTED FROM BAMBOO STALKS THAT GROW WILD IN SHALLOW MUD FLATS ALONG THE COAST. CULTIVATED NORI IS HARVESTED BY MACHINES, MECHANICALLY SHREDDED AND DRIED BY BLOWERS. THE JAPANESE EAT OVER NINE BILLION SHEETS A YEAR.

NUTRIENTS: NORI IS AN ABUNDANT SOURCE OF CALCIUM, POTASSIUM, MANGANESE, MAGNESIUM, AND PHOSPHORUS. IT IS HIGH IN VITAMIN A, HAS TEN TIMES THE NIACIN OF SPINACH, AND AS MUCH VITAMIN C AS TOMATOES.

PREPARATION: TOAST NORI THAT HAS NOT BEEN PRE-TOASTED FOR SUSHI MAKING OR CRUMBLING. SUSHI-NORI IS A PREMIUM QUALITY THAT IS PRE-TOASTED. THE SHEETS ARE UNBLEMISHED AND EASY TO ROLL. NORI CANNOT BE FROZEN AS IT BECOMES INEDIBLE WHEN THAWED.

USES: NORI IS USED MOST WIDELY FOR SUSHI AND FOR WRAPPING FOODS AND CONFECTIONS. IT CAN BE CUT INTO SQUARES AND FRIED; CHOPPED OR BROKEN, IT CAN BE ADDED TO SOUPS, SALADS, CASSEROLES, AND STEWS; TOASTED AND CRUMBLED, IT CAN BE USED AS A GARNISH.

TOASTED NORI

TAKE TWO SHEETS OF NORI AND PUT THEM TOGETHER WITH THE SHINY SIDES TOUCHING. GENTLY, BUT QUICKLY, PASS THE SHEETS OVER A MEDIUM GAS FLAME. CONTINUE UNTIL THE NORI CHANGES TO AN IRIDESCENT GREEN. THE TEXTURE WILL CHANGE FROM SMOOTH TO COARSE-LOOKING. REPEAT AGAIN WITH TWO MORE SHEETS.

UNPROCESSED NORI DOES NOT COME IN SHEETS. IT CAN BE ROASTED IN A DRY, CAST-IRON SKILLET OVER MEDIUM HEAT. TURN OFTEN AND FEEL FOR CRISPNESS WITH FINGERS. ANOTHER METHOD IS TO BAKE NORI IN A 200° OVEN UNTIL CRISP. WATCH IT CLOSELY, AS IT BURNS EASILY.

CRUSH TOASTED NORI IN YOUR FINGERS, WITH A MORTAR AND PESTLE, OR BETWEEN LAYERS OF WAX PAPER, TO MAKE A FINE POWDER TO USE AS A CONDIMENT. CUT TOASTED NORI INTO 2" SQUARES TO USE IN SNACKS OR AS "CHIPS". USE TOASTED NORI POWDER OR CHIPS AS A GARNISH ON RICE, SOUPS, CASSEROLES, SALADS, SANDWICHES, QUICHES, FRITATAS, AND VEGETABLES.

SUSHI RICE

SUSHI IS A COLORFUL AND FLAVORFUL ROLL OF FILLED RICE THAT IS WRAPPED IN NORI. TRADITIONALLY IT IS PREPARED WITH VINEGARED RICE, FISH, AND VEGETABLES AND SERVED AS AMERICANS WOULD SERVE SANDWICHES. A SUSHI PARTY COULD BE A LOT OF FUN WITH PEOPLE MAKING THEIR OWN FROM A WIDE CHOICE OF INGREDIENTS.

RICE FOR SUSHI

1 CUP PEARL RICE
1 STRIP KOMBU, 1" LONG
1 1/4 CUPS WATER
2 TBSP. RICE VINEGAR
1/2 TEASPOON SALT
2 TEASPOONS HONEY

WASH THE RICE WELL AND DRAIN. COMBINE KOMBU AND WATER IN A SAUCEPAN AND BRING TO A SIMMER, REMOVING KOMBU JUST AS THE BOILING POINT IS REACHED. ADD THE RICE, COVER, AND BRING TO A BOIL OVER HIGH HEAT. REDUCE TO LOW AND STEAM FOR 15 MINUTES. REMOVE FROM HEAT AND LET STAND FOR 10 MINUTES MORE. COMBINE VINEGAR, SALT, AND HONEY IN A SMALL SAUCEPAN AND HEAT SLIGHTLY TO MELT HONEY. SPREAD RICE IN A WOODEN BOWL OR ON A FLAT PLATTER AND SPRINKLE WITH VINEGAR MIXTURE. MIX WITH A FLAT SPATULA UNTIL IT STOPS STEAMING. IT IS NOW READY TO BE USED FOR SUSHI.

SUSHI FILLINGS

FILLINGS:

1. 1/2 CAKE VERY FIRM TOFU
   1 TEASPOON CAPERS
   2 TEASPOONS CHOPPED PIMIENTO
   1 TEASPOON CURRY POWDER

MASH THE TOFU AND MIX IN CAPERS, PIMIENTO, AND CURRY POWDER. TASTE AND ADD SEASONINGS TO TASTE.

2. 2 EGGS
   1 TBSP. WATER
   1 TEASPOON CHOPPED CHIVES OR GREEN ONION

BEAT EGGS WITH WATER AND CHIVES. POUR A LITTLE OIL IN A SKILLET, ENOUGH TO COVER THE BOTTOM AND COOK EGGS GENTLY, OVER LOW HEAT TO MAKE AN OMELET. CUT INTO LONG STRIPS WHEN COOLED.

3. 5 DRIED SHIITAKE MUSHROOMS
   ~ WATER TO COVER
   2 TEASPOONS TAMARI
   2 TEASPOONS SHERRY
   1/4 TEASPOON ONION POWDER

SOFTEN MUSHROOMS IN WATER FOR SEVERAL MINUTES. SIMMER THEM IN 1/4 CUP OF THE SOAKING LIQUID WITH THE TAMARI, SHERRY, AND ONION POWDER. COOL, DRAIN, AND CHOP VERY FINE.

SUSHI FILLINGS, CONTINUED

4.
- 4 CARROTS
- 6 GREEN ONIONS
- 2 RIBS (STALKS) CELERY
- 2 ZUCCHINI SQUASH
- 2 TEASPOONS TAMARI

CUT VEGETABLES INTO LONG STRIPS. STEAM IN A LARGE FRYING PAN WITH THE TAMARI AND A SMALL AMOUNT OF WATER UNTIL JUST TENDER, BUT NOT SOFT.

5. SHRIMP, CRABMEAT, TUNA, OR ANY FISH OR SEAFOOD YOU LIKE, COOKED AND FLAVORED TO YOUR TASTE.

6. RAW AND FRESH: MUSHROOMS, CUCUMBER, ZUCCHINI, SCALLIONS, CHIVES, OR OTHER VEGETABLES SLICED LONG AND THIN.

7. CHOPPED AND SLIGHTLY COOKED FRESH SPINACH.

8. SMOKED OR PICKLED SALMON, HERRING, OR OTHER FISH.

CONDIMENTS:

FLAVORED VINEGAR, TAMARI, TOASTED SESAME SEEDS, GRATED HORSERADISH, SHREDDED DAIKON, AND MUSTARDS - HOT, SWEET, AND DRY.

ASSEMBLING SUSHI

ASSEMBLING SUSHI

YOU WILL NEED:
- VINEGARED RICE
- FILLINGS
- CONDIMENTS
- TOASTED NORI SHEETS
- FLAT CUTTING BOARD
- A BAMBOO MAT OR WAX OR PARCHMENT PAPER 9" X 9"

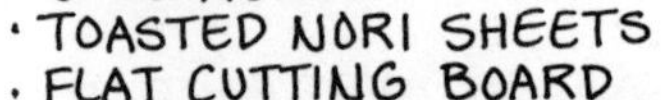

PLACE A SHEET OF TOASTED NORI (SEE PAGE 48) ON THE MAT OR PAPER. SPREAD OVER IT A 1/2" LAYER OF RICE, LEAVING A MARGIN ALL AROUND THE SIDES OF 3/4 OF AN INCH. (HINT: SPREAD THE RICE WITH MOISTENED HANDS.) LAY STRIPS OF FILLING IN ANY COMBINATION YOU CHOOSE DOWN THE CENTER, HORIZONTALLY. LIFT THE NEAR SIDE OF THE MAT OR PAPER AND ROLL TOWARD THE FAR SIDE, GENTLY PRESSING THE ROLL TO MAKE IT FIRM. PRESS THE ENDS WITH YOUR FINGERS TO SEAL THE SUSHI ROLL. REMOVE THE MAT OR PAPER AND CUT THE ROLL INTO 1/2" TO 1" SEGMENTS. ARRANGE ON A PLATTER AND REPEAT. GARNISH THE PLATTER WITH SLICED FRESH VEGETABLES AND SERVE AS AN APPETIZER OR SNACK WITH COCKTAILS, WARM SAKÉ OR BEER.

NORI CONDIMENT

ROAST 2 OUNCES OF DRY NORI IN A 200° OVEN FOR 5 MINUTES AND GRIND WITH 1/3 CUP SESAME SEEDS WITH A MORTAR AND PESTLE, (SURIBACHI). USE AS A CONDIMENT WITH GRAINS OR FISH. STORE IN THE REFRIGERATOR.

NORI POWDER

SPREAD 3 TO 6 OUNCES NORI ON A COOKIE SHEET AND ROAST IN A 250° OVEN FOR ABOUT 10 MINUTES. CRUSH OR GRIND INTO A POWDER IN A BLENDER OR MORTAR AND PESTLE (SURIBACHI). STORE IN A TIGHTLY CLOSED JAR IN THE REFRIGERATOR. ADD TO SOUPS, SALAD DRESSING, CASSEROLES, AND BAKED GOODS FOR A DELIGHTFUL FLAVOR AND ADDED NUTRITION.

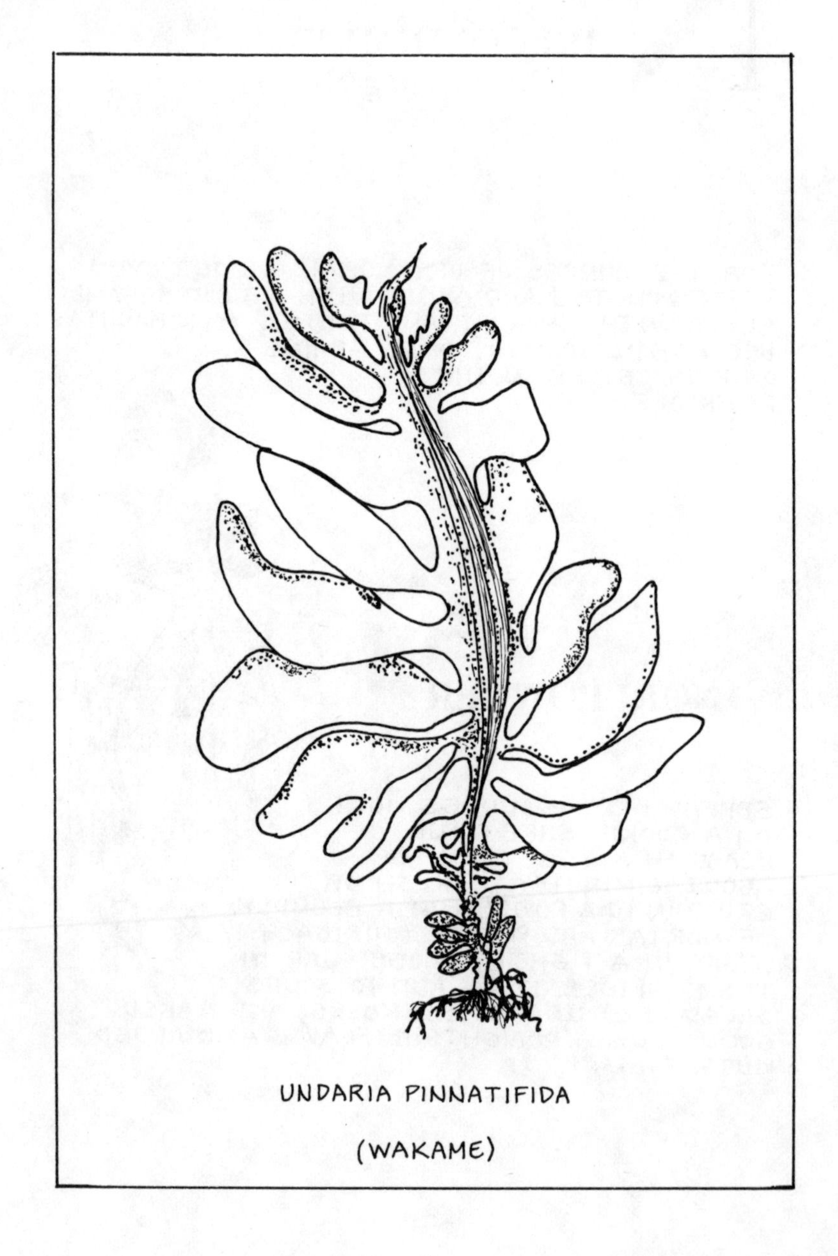

UNDARIA PINNATIFIDA

(WAKAME)

WAKAME

WAKAME IS SO IMPORTANT TO JAPANESE DIET, THAT INSTEAD OF HARVESTING IT FROM THE SEABED WITH LONG POLES, AS WAS DONE FOR CENTURIES, IT IS NOW EXTENSIVELY CULTIVATED. THE WILD PLANTS ARE STILL HIGHLY PRIZED BECAUSE THEY RETAIN THEIR CRISPNESS IN HOT LIQUIDS. THE CULTIVATED PLANTS TEND TO SOFTEN.

WAKAME MEANS "YOUNG GIRL." THE TWO TO FOUR FOOT LONG BLADES ARE WASHED, CUT DOWN THE MIDDLE, AND HUNG UP TO DRY IN THE SUN. ANOTHER WAY IS TO PRESS THEM INTO THICK SHEETS AND DRY THEM FLAT ON THE GROUND.

THE LEAVES HAVE A SWEET TASTE AND ARE USED EXTENSIVELY IN MISO SOUP. WHEN SOAKED, WAKAME BECOMES AN ATTRACTIVE GREEN COLOR.

<u>NUTRIENTS</u>: WAKAME IS A SUPERIOR SOURCE OF CALCIUM AND NIACIN. IT IS A GOOD SOURCE OF PROTEIN, IRON, SODIUM, AND MANGANESE, AS WELL AS VITAMINS A, B_1, B_2, AND C.

<u>PREPARATION</u>: TO RECONSTITUTE WAKAME, SOAK IT UNTIL SOFT, USUALLY 10 TO 20 MINUTES; DRAIN AND SAVE THE SOAKING WATER FOR SOUP. WAKAME WILL SWELL TO ABOUT 5 TIMES ITS DRY SIZE. AFTER SOAKING, CUT OUT THE CENTER RIB. CHOP THE RIB VERY FINE AND CUT THE SOFT PART OF THE LEAF INTO BITE SIZED PIECES.

<u>USES</u>: ADD WAKAME TO SOUPS, SALADS, CASSEROLES, AND STEWS. IT CAN BE ROASTED AND DEEP FRIED, OR EATEN FRESH AS A VEGETABLE. IT IS ALSO USED PICKLED AND MARINATED AS A CONDIMENT.

MISO AND WAKAME SOUP

SERVES 6

1 TO 2 OUNCES WAKAME (DRY)
6 CUPS WATER
2 TBSP. MISO
1/4 TO 1/2 CAKE FIRM TOFU
5 OR 6 THINLY SLICED SCALLIONS

SOAK THE WAKAME FOR ABOUT 15 MINUTES. DRAIN AND SAVE THE WATER, ADDING TO IT TO MAKE 6 CUPS. CUT THE RIB FROM THE WAKAME LEAVES AND SAVE FOR ANOTHER DISH. CHOP THE SOFT LEAF PARTS INTO 1" PIECES. BRING THE STOCK/WATER TO A BOIL. ADD THE MISO AND STIR UNTIL COMPLETELY DISSOLVED, THEN ADD THE TOFU AND WAKAME. WHEN THE SOUP COMES TO A BOIL AGAIN, SPRINKLE WITH SCALLIONS AND SERVE AT ONCE.

SWEET AND SOUR WAKAME

SERVES 6

- 2 OZ. DRIED WAKAME
- 8 OZ. TEMPEH, CUBED
- 3 TBSP. CORNSTARCH
- 1 1/2 CUPS APPLE JUICE, ORANGE JUICE, OR WATER
- 1/3 CUP TAMARI
- 4 TBSP. HONEY
- 6 TBSP. VINEGAR
- 3 TBSP. SESAME OR PEANUT OIL
- 1 LARGE ONION, CUT IN WEDGES
- 1 LARGE ONION, CUT IN WEDGES
- 1 LARGE CARROT, SLICED
- 1 GREEN PEPPER, CUT IN PIECES
- 1 PEELED APPLE, CUT IN PIECES
- 4 THIN SLICES GINGER ROOT, MINCED
- 1/2 CUP ROASTED NUTS OR SEEDS

SOAK THE WAKAME FOR 15 MINUTES. DRAIN AND SAVE THE WATER. CUT AWAY THE MID RIB AND CUT THE LEAVES INTO 1" PIECES. SET ASIDE. CUT THE TEMPEH INTO BITE SIZED CUBES AND STEAM OVER THE WAKAME WATER FOR 10 MINUTES. SET ASIDE.

TO MAKE THE SAUCE, DISSOLVE THE CORNSTARCH IN THE JUICE; ADD THE TAMARI, HONEY, AND VINEGAR AND MIX WELL. HEAT THE OIL IN A WOK OR LARGE SKILLET AND STIR-FRY THE ONION, CARROT, GREEN PEPPER, APPLE, AND GINGER ROOT UNTIL COOKED TO YOUR TASTE. ADD THE TEMPEH AND WAKAME AND STIR FRY FOR ANOTHER FEW MINUTES. ADD THE SWEET AND SOUR SAUCE AND COOK UNTIL SAUCE THICKENS. POUR ONTO A SERVING PLATTER AND SPRINKLE WITH NUTS. SERVE WITH STEAMED RICE.

SOME COMMON SEA GREENS

SPECIES	USE	COMMON NAME	WHERE FOUND
Brown algae (Phaeophyta)			
		HIJIKI	
Hizikia fusiforme	food	Hijiki	Temperate Japan,
Sargassum echinocarpum	food	Limu-kala	Hawaii, Taiwan, Indian Ocean
		KOMBU	
Laminaria japonica	food	Kombu	Japan
Laminaria sinclairii	food	Kelp, tangle	Mid Pacific
Laminaria andersonii	algin	Split whip wrack	No. Pacific
Laminaria digitata	food	Horsetail kelp	Atlantic coasts
Macrocystis pyrifera	algin, food	Giant kelp	Cold seas world-wide
		WAKAME	
Alaria esculenta	food, algin	Winged kelp, Honey ware	Atlantic and Pacific coasts
Undaria pinnatifida	food	Wakame	North coasts of Japan, U.S., British Isles
		ARAME	
Eisenia Arborea	food, algin	Arame, Sea Oak	Japan, Pacific coasts, So. America
Red algae (Rhodophyta)			
		NORI	
Porphyra subordiculata	food	Nori	Japan, California
Porphyra crispata	food	Nori	Hawaii, Europe
Porphyra dentata	food	Nori	Phillipines
Porphyra tenera	food	Asakusa-nori	Cultivated in Japan
Porphyra umbilicalia	food	Purple laver	Colder waters
Porphyra perforata	food	Red laver-nori	Pacific & Atlantic

SOME COMMON SEA GREENS

SPECIES	USE	COMMON NAME	WHERE FOUND
Red algae (Rhodophyta)			
	AGAR or AGAR-AGAR		
Gelidium amansii	agar	Kanten, Tengusa	Temperate Japan
Gelidium latifolium	agar		Atlantic coasts
Gelidium cartilagineum	agar		Atlantic and Pacific coasts
Chondrus crispus	carrageen	Irish Moss	Cold waters Atlantic
Gigartina exasperata	carrageen food	Turkish towel	No. California to Canada
Ascophyllum nodosum	algin	Rockweed	Atlantic, colder regions
	DULSE		
Palmaria palmatta	food used widely in Europe & U.S.	Dulse Soul-söll	Cold waters, world-wide.
Green Algae (Chlorophyta)			
Ulva lactuca	food	Sea lettuce Green Laver	Worldwide in cold and temperate seas
Enteromorpha linza	food	Sea lettuce Green laver	World wide.
Blue-Green Algae			
Nostoc sp.	food	Spirulina	Fresh water, mostly cultivated

SEA GREEN SOURCES

American Sea Vegetable Company John Olson P.O. Box 773 Vashon Island, Washington 98070 (206) 622-6448 Or 463-3834	Cultivating and producing Nori and Wakame
Maine Coast Sea Vegetables Shepard and Linnette Erhart Shore Road Franklin, Maine 04634 (207) 565-2907	Dulse, Alaria, Kombu and Nori available.
Mendocino Sea Vegetables John and Eleanor Lewallen Box 372 Navarro, CA 95463 (707) 895-3741	Mendocino Nori, Mendocino Wakame, Neptune's Delight, Sea Palm. Seasonal availability of Mendocino Dulse; other sea vege- tables upon request. Packaged & bulk prices. Shelf dis- play available.
New England Sea Vegetables Larch & Jan Hanson Box 15 Steuben, Maine 04680 (207) 546-2875	Kelp, Alaria, Dulse, Horsetail Kelp, Sea Lettuce, Irish Moss. Harvesting & recipe book available. Wholesale & retail prices.
Ocean Energy Mathew P. Hodel 4421 Soquel Ave. Soquel, CA 95073 (408) 476-4265	Red Nori, Dulse, Nereo Kelp, Sea Palm, Alaria, Sweet Nereo Pickles. Wholesale & retail prices.
Oregon Sea Vegetables Evelyn McConnaughey 1653 Fairmont Blvd. Eugene, OR 97403 (503) 345-0227	Pelvetiopsis, Alaria "Oregon Wakame" Laminaria "Kombu," Porphyra "Nori." 2 sea vegetable recipe books available.
Rising Tide Sea Vegetables Kate Marianchild P.O. Box 228 Philo, CA 95466 (707) 895-3278	Nori, Wakame. Dulse, Sea Palm, Ulva available upon request. Wholesale and retail prices.

BIBLIOGRAPHY

1. Arasaki, Seibin, D. Agr. and Teruko Arasaki, D. Sc. 1983. *Vegetables From The Sea.* Japan Publications, Inc. Tokyo, Japan.
2. Ford, Richard. 1980. *The Soy Foodery Cookbook: Tofu, Tempeh, Miso, Sea Vegetable Recipes & Resources,* 1980, Day & Night Graphics, Santa Barbara, CA.
3. "Growing Interest in Oriental Foods Livens Interest for Market of Sea Vegetables," Robin Hansen, *National Fisherman,* Nov., 1982, p. 58.
4. Hanson, Larch. 1980. *Edible Sea Vegetables of the New England Coast.* Private printing. Steuben, Maine.
5. "Kelp and other gifts from Neptune's Garden," Dominick Bosso, *Prevention,* Vol. 10, p. 95.
6. "Learning to Love Seaweed;" Martha Wagner, *New Age Magazine,* March 1980, pps. 75-77.
7. Lewallen, Eleanor & John. 1983 *The Sea Vegetable Gourmet Cookbook and Foragers Guide.* P.O. Box 372, Navarro, CA 95463.
8. Madlener, J.C. 1977. *The Sea Vegetable Book.* Clarkson N. Potter, Inc., New York.
9. Madlener, J.C. 1977. *The Sea Vegetable Book: Foraging and Cooking Seaweed.* Clarkson N. Potter Inc., New York.
10. Madlener, J.C. 1981. *The Sea Vegetable Gelatin Cookbook and Field Guide.* Woodbridge Press.
11. McConnaughey, Evelyn. 1980. *Sea Vegetable Recipes from the Oregon Coast.* Oregon Institute of Marine Biology, Charleston, Oregon 97420.
12. Turner, Nancy J. 1975. *Food Plants of British Columbia Indians.* Part I. Coastal Peoples Handbook. No. 34, B.C. Provincial Museum, Victoria, British Columbia.
13. "Vegetables from the Deep," Sharon Ann Rhoads, *EW Journal,* April, 1979, pps. 68-71.

INDEX

We dedicate this book to
Jan Ingram
and
Cousin Jim

. . . with fronds like these, who needs anemones?